Hilton Head Island, South Carolina: Naturalists call the waters here "Nature's Soup," they are rich with nutrients allowing sea life to thrive. The ecosystem here is a nursery for sea creatures. Tides wash in Spartina Grass to shelter ghost crabs that burrow in the sand. Their holes catch seeds, beginning the process of building dunes with glistening sea oats.

Each May to October the beach becomes a sea turtle nursery. Mother loggerhead turtles return to the beach where they were born. They make their way to the dunes during the night to dig nests and leave eggs to mature in the warm sun. Two months later hatchlings scamper from the nests heading back to the sea to begin the adventure that is their life.

Hilton Head has an active sea turtle protection project that locals and guests of the island can join. You can adopt a nest for the summer and get regular reports on your hatchlings. You can also help while you are on the island; just walk on the beach with a trash bag to pick up small objects that could cause a mother or her hatchling serious damage. Small pieces of plastic, broken up styrofoam, straws, sand bucket handles, plastic shovels, balloons, pieces of spent fireworks, all of these and more are dangerous to the wildlife. In the evenings the Turtle Team can be seen carrying shovels or rakes to smooth the path from nests to the ocean for hatchlings that might emerge during the night. New nests are marked each morning.

If you come out at sunrise you may see tracks where a mother turtle came up to nest during the night. We must also protect her pathway. Because so many people come for vacation and don't know about the turtles they may not know how important it is to:

"Put The Beach To Sleep At Night"

Here is where we need you to teach your parents, grandparents and your brothers and sisters. Sea turtles exist around the world in a band from Canada to Brazil, so no matter where you come from, or where you go in the future your pledge to be on the Turtle Patrol can be a lifetime commitment.

Please enjoy the beauty that nature created for you, and remember the future is yours to protect. Now enjoy the story of Elizabeth the Loggerhead.

This story is dedicated to David, Kristin, Stephen and Brian, my hatchlings!

© 2017 Elizabeth Belenchia

Published by Eco Planet Publishing, Inc.

All Rights Reserved

ISBN 978-0-9995760-0-7

Disclaimer:
The illustrations in this book are artistic interpretations
of our story from the artists perspective.

My Walk
to the Water
A Mother Turtle Speaks

Elizabeth Belenchia

ILLUSTRATED BY:
Ryan Collins

LAYOUT & DESIGN BY:
Brian Carroll
Marine Marketing Group, Inc.

PUBLISHED BY:

ECO PLANET
PUBLISHING, INC

We donate to "1% for the Planet"

Hello, my name is Elizabeth.
My family is older than the dinosaurs.
I spend my days swimming long distances looking for food because I am very big and swimming takes a lot of energy. I eat jellyfish and blue crabs and like the crunch of conch that I crack with my beak and strong jaw muscles.

I am a loggerhead sea turtle!

Sometimes I have a seaweed salad, it is refreshing. I have a very large head and four strong flippers which propel me through the ocean.

Many school children went to the government to share our beauty and the threats we experience daily that could lead to our extinction. As a result I was named the state reptile of South Carolina, the place I call home.

I was born thirty years ago on the Carolina barrier island called Hilton Head. I hatched in the middle of the night. I could see the Carolina moon in the sky and its beams led me to the warm water gently crashing on the sand where I was born.

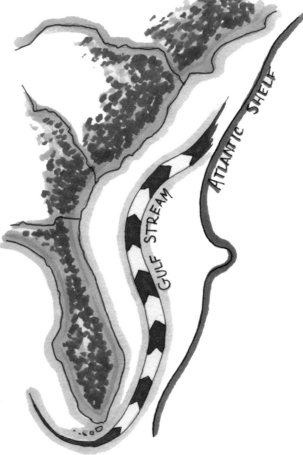

They say I have a built-in GPS system that brings me back to this shoreline to nest and have my own children.

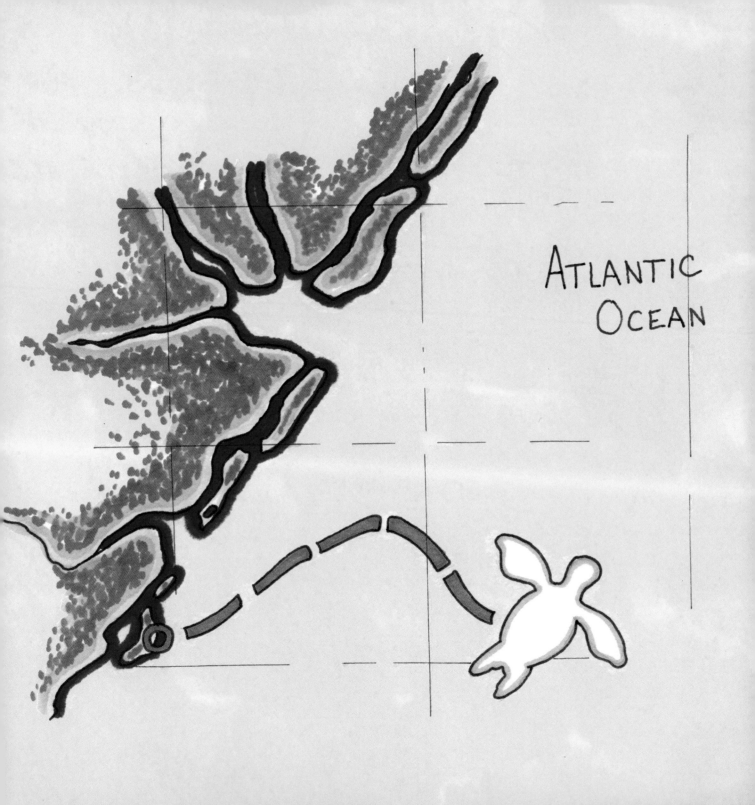

ATLANTIC
OCEAN

I have a rusty-brown, heart shaped carapace
(that's my top shell) and a yellow plastron
(my bottom shell).
I now weigh 250 pounds.

I can swim fifteen miles per hour
with my featherweight
bones and secret air
pockets. I swim to
Florida sometimes and
once even went to
the Azores and Africa
on my way back to
Hilton Head Island.

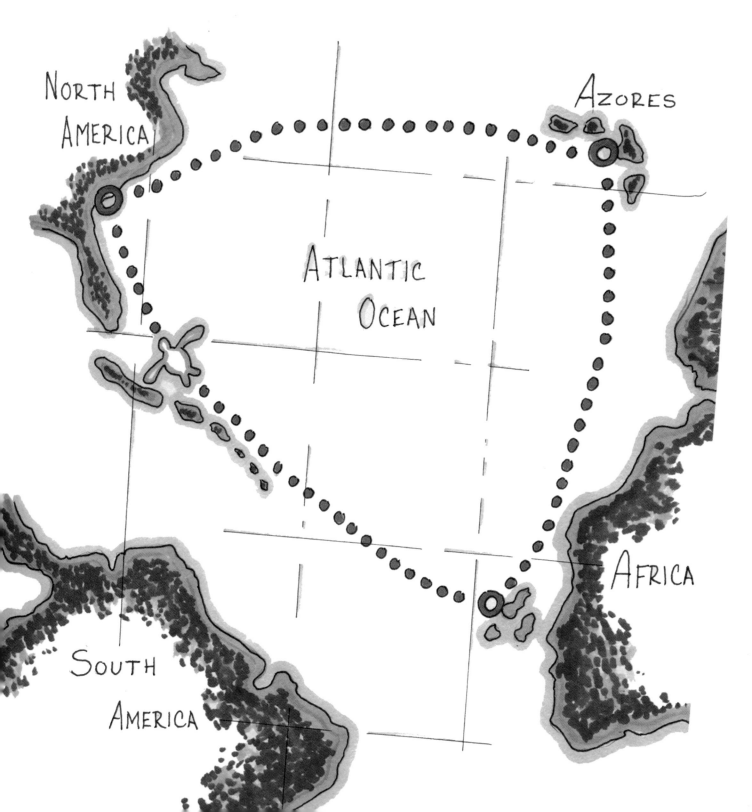

NORTH
AMERICA

AZORES

ATLANTIC
OCEAN

AFRICA

SOUTH
AMERICA

I like the shelf along the coast that has ripples in the sand. I have to come up for air so the shallow water is a great place for me to play and see other animals swimming around. In winter I slow down and am able to go into deeper water because I can hold my breath for long periods of time when I am not swimming.

PORT
ROYAL
SOUND

This spring I met a male turtle while swimming in Port Royal Sound. He nuzzled my nose to let me know I was special. Port Royal is a beautiful place to swim with wonderful water. It is full of good food and lots of friends who like to play. We decided it was time for us to have a family.

A turtle family is very different.
Most of our life is spent in the water.
My mother crawled out of the
water on a warm summer night
in June. She had to pull herself
up on the beach above the
highest tide line and dig a pit
with her back flippers.
She left tracks in the sand.

Then she gently dropped me deep in the hole she dug in the dune called a nest with 120 other soft eggs nearly the size of golf balls. A nest can be four feet long, which is almost as tall as you are! It took three hours for her to drop all of the eggs into her nest. If you saw her you might think she was crying but it's just salt coming from her eyes. She swept sand over us with her back flippers to protect us from raccoons, dogs, ghost crabs, birds, ants and people who might harm us.

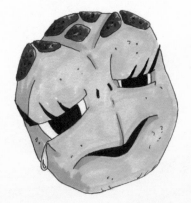

She slowly went back to the ocean because she was very tired and needed to float in the water. She stayed on the ocean shelf to rest and recover so she could lay her next family of eggs. She can make three to six nests over the summer, one every two weeks.

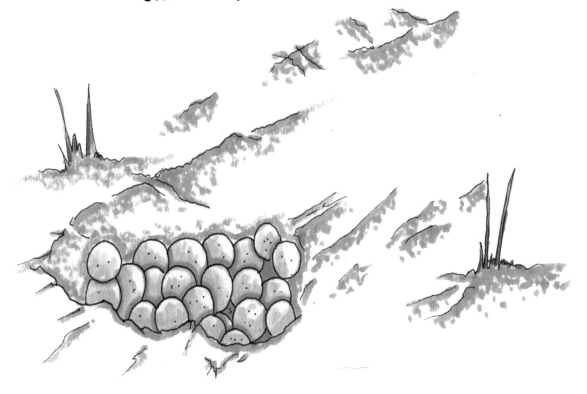

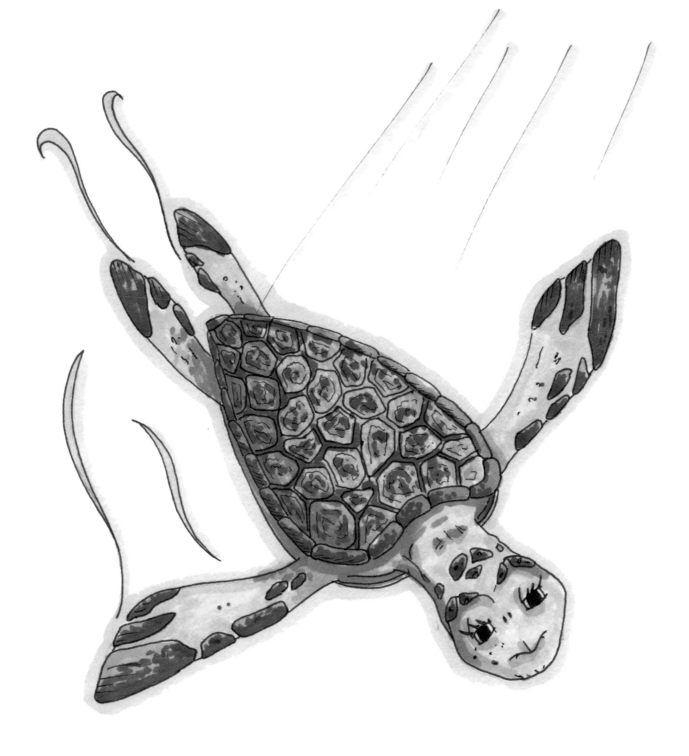

Over time we all grew inside of our shells.
We enjoyed feeling the warm sun, rain and cool
temperatures at night. We loved the rain because
of the splashing sound it made on our nest.
The rain cooled the sand and our shells. This was
especially important because the cool nights meant
we wouldn't just have sisters but also brothers too.
When it is too hot we only have sisters and that
doesn't help our families continue to grow.

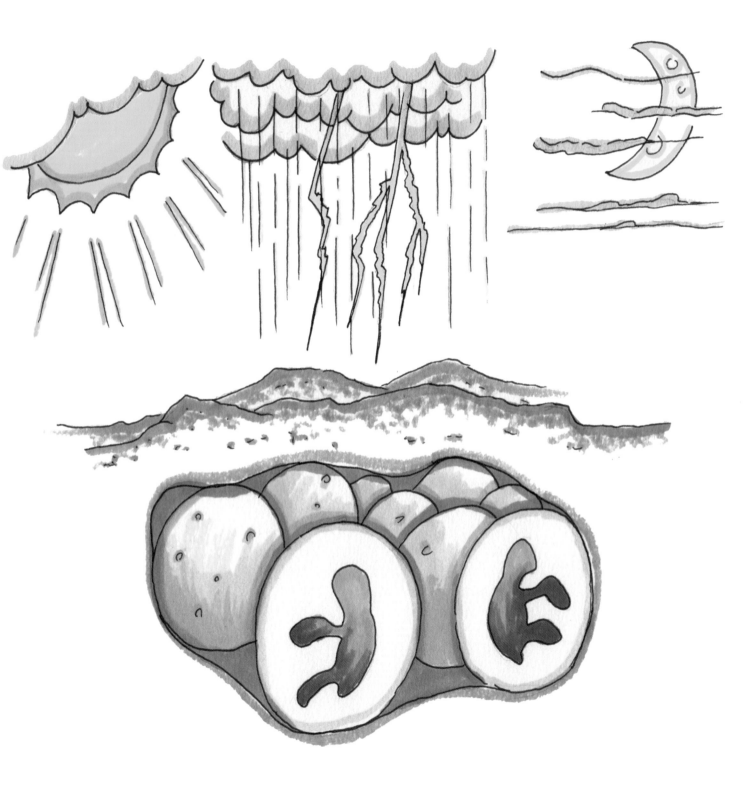

For two months we stretched and grew,
A carapace formed on our backs and little
flippers came out. We all had large heads,
which we would need to crush shells and eat
the juicy seafood as we grew. We were
now only two inches long! Then we said
"These eggs are not big enough to hold us any longer."
We began to peck from the inside with our beaks
and our shells began to crack. We wiggled and
twisted until they opened. We pushed the shells
to the bottom of the nest so we could stand on
them and made sure the whole family was free.

We saw our brothers and sisters and we somehow knew we had to get into the water before the sun woke up.

Now only two inches long, we went
scampering toward the sea.
We sensed there was danger.

Our walk to the water can be very dangerous
with lots of obstacles. We might fall
into holes in the beach like a moat children
dug around a beautiful sand castle.
A ghost crab or seagull might grab us
and try to eat us.

We can mistake house lights
for the reflections of the ocean at
night. To help us you can turn off lights
and cover your flashlight with a red
turtle filter. If we crawl towards land
and not the water, we could get lost in the
sand dune and may not survive.

A hurricane could bring
ferocious wind and waves. We would
have to fight to get to the ocean.
Those plastic shovels and bags would be
obstacles that could now hit us on the
way to the safety of the ocean.

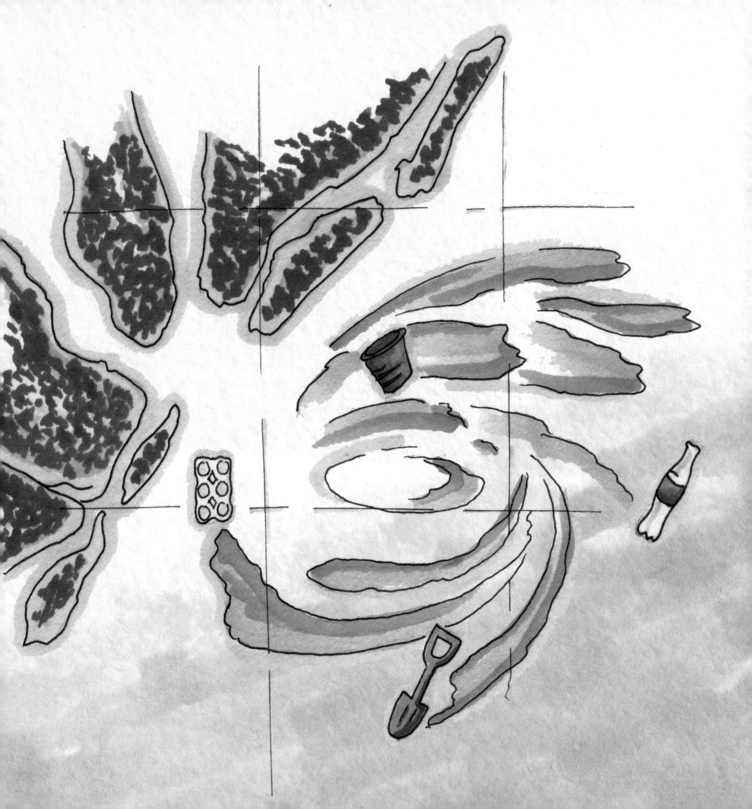

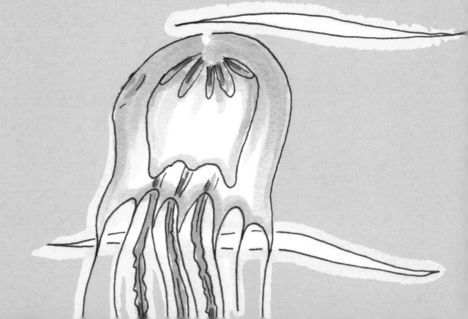

Sea turtle families are different than your family. Only one out of a thousand of us will survive. We have to be strong and get to the sea. If we are lucky and survive we could live to be 100 years old! Mother didn't have time to warn us about plastic floating in the water that shines like delicious jellyfish.

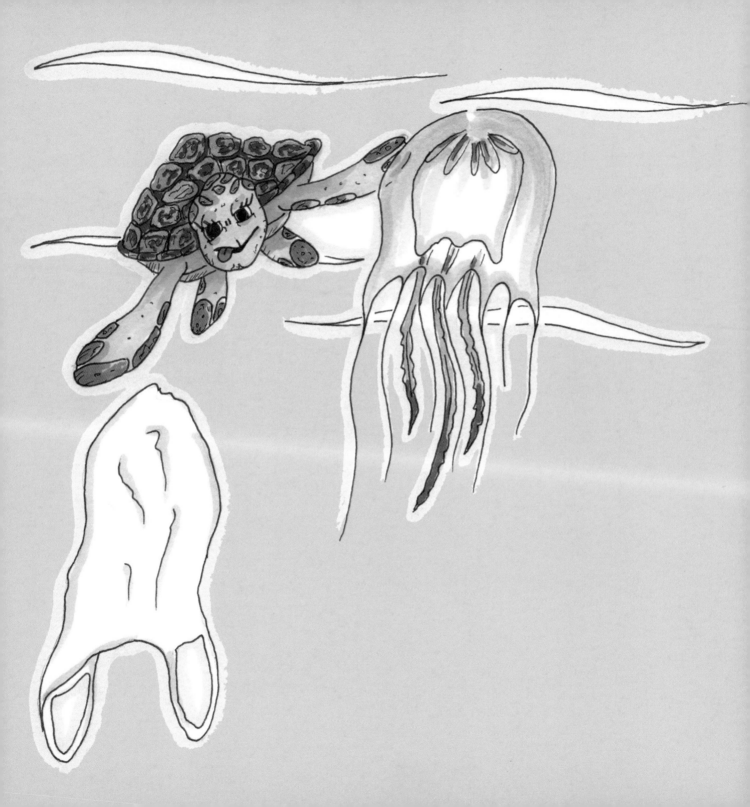

My GPS is just developing, I am a baby and I need your love and protection. You are much bigger than me, but one day I could be over 400 pounds. When I grow older I will return to Hilton Head and see your children enjoying the island because of the care and protection you gave me many years ago. I will always trust your children and their families to care for my turtle family.

I am like every child in the world.
I want to live and enjoy this beautiful
beach of Hilton Head Island where
my family has been coming
for more than a thousand years.
I love this precious Atlantic Ocean
and all the good food it
holds for me and my family.

Thank you for taking care of
my home and do have fun on
the beach while I am growing
and swimming in the sea. I
hope we will see each other
again one day as we share this world.
In the meantime before you go to bed

"Please Put The Beach To Sleep"

Fill holes in the sand, pick up plastic
and trash from the shore, remember
no lights, fires or fireworks after ten o'clock at
night and always watch turtles from a distance.

Please help make our journey safe.